Bibliographic information published by the German National Library:

The German National Library lists this publication in the National Bibliography; detailed bibliographic data are available on the Internet at http://dnb.dnb.de .

Imprint:

Copyright © 2013 GRIN Verlag, Open Publishing GmbH
Print and binding: Books on Demand GmbH, Norderstedt Germany
ISBN: 978-3-668-03150-0

This book at GRIN:

http://www.grin.com/en/e-book/305240/visualization-of-large-unstructured-grids-efficient-point-and-cell-location

Philipp Güth

Visualization of Large Unstructured Grids. Efficient Point and Cell Location

GRIN Publishing

GRIN - Your knowledge has value

Since its foundation in 1998, GRIN has specialized in publishing academic texts by students, college teachers and other academics as e-book and printed book. The website www.grin.com is an ideal platform for presenting term papers, final papers, scientific essays, dissertations and specialist books.

Visit us on the internet:

http://www.grin.com/

http://www.facebook.com/grincom

http://www.twitter.com/grin_com

Visualization of Large Unstructured Grids

Efficient Point and Cell Location

Philipp Güth

June 22, 2013

Computer Graphics and Visualization
Interdisciplinary Center for Scientific Computing (IWR)
Heidelberg University

Abstract - Visualization of large datasets, especially the visualization of unstructured grids, is a challenge due to the unstructured nature of the data which oftentimes causes large overheads in memory as well as performance problems on large grids. Problems emerge because existing solutions generally presuppose properties like uniform point distributions for datasets which are usually not existent in unstructured grids. These issues become particularly problematic on large grids since the existing solutions, if they work at all for unstructured grids, do not scale well. In this paper I will present two innovative approaches to visualization in large, unstructured grids. The first approach was developed by Max Langbein, Gerik Scheuermann and Xavier Tricoche. It makes use of cell adjacency and a complete adaptive k-d tree and utilizes ray shooting to locate points for visualization. The second approach was developed by Christoph Garth and Kenneth I. Joy. They use an innovative data structure, the *celltree* which is based on a bounding interval hierarchy, in order to narrow down the number of cells that conceivably contain points for visualization. Both approaches present memory-efficient and performant solutions for visualizing large unstructured grids, the approach of Garth and Joy further focuses on numerical robustness. The main difference between the two papers is that the work of Garth and Joy designs a data structure based on points and attempts to narrow down the number of cell candidates and subsequently performs a simple check for inclusion, whereas in the work of Langbein et al. the data structure design is based on the cells and uses ray tracing after making an educated guess for a cell close to the searched point. In other words, Garth and Joy present an approach to cell location, Langbein et al. present an approach for point location.

1 Introduction

While grids can oftentimes be visualized quite easily when assuming preconditions for their structure, in practice large unstructured grids are prevalent in many scientific fields and industrial applications such as wind flow analysis or structural mechanics. Being able to perform visualization tasks in reasonable time is crucial for these applications, yet the transition from a structured grid to an unstructured one is far from trivial. The visualization of unstructured grids is a challenging endeavor that is based on a variety of spatial subdivision techniques and methods for point localization. Therefore I will introduce the underlying data structures and localization methods first to establish a solid foundation for the large field of point and cell location methods.

Afterwards I will present the works of Max Langbein, Gerik Scheuermann and Xavier Trichoche on point location in large unstructured grids ([4]) and the work of Christoph Garth and Kenneth I. Joy on cell location in unstructured grids ([3]). The approach of Langbein et al. was presented in 2003 and provided a remarkable gain in performance both in regards to memory overhead as well as performance. Garth and Joy base some of their ideas on the work of Langbein et al., however they use a different approach to storing data in order to increase performance even further and reduce memory overhead even more.

2 Foundations

In order to understand how and, more importantly, why the approaches introduced in this paper work efficiently, it is necessary to understand the underlying data structures and concepts. The point location method of Langbein et al. uses the properties of an adaptive k-d tree whereas the the cell location method of Garth and Joy utilizes the advantages of a bounding interval hierarchy data structure, both of which are introduced in the following chapter.

2.1 k-d trees

The idea of utilizing the structure of multidimensional binary search trees for storing information in order to perform associative searches on that data structure was first introduced by Jon Louis Bentley at Stanford University in 1975 with his definition of the k-d tree. A k-d tree is essentially a k-dimensional binary tree. That means it is a binary tree, in which every node has the dimension k. Speaking in a more graphical way it is a tree, in which every node is a hyperplane that subdivides a k+1-dimensional space. The children of each node represent splitting planes that further subdivide space on both sides of the hyperplane. For further information on k-d trees see the work of J.L. Bentley [5].
When used for point visualization, k-d trees are usually constructed by placing splitting planes through the points contained in the structure to visualize. An illustration of the concept is shown in figure 1. It shows a 2-dimensional plane that contains points. In order to store the points a 1-dimensional k-d tree is created by using 1-dimensional splitting planes to subdivide the plane, i.e. straight lines that subdivide the space of the plane. Note that the k-d tree subdivides space, not the grid itself. Therefore cells of a grid can intersect with more than one leaf of a tree that partitions space over

the grid, however every leaf does only contain one vertex of the grid. For further deliberations on construction and usage of k-d trees see [1].

2.2 Bounding Volume Hierarchy

A common problem with ray-tracing is that the objects in a given space have complex shapes. An approach to better performance both in regards to memory usage as well as computational speed is to construct bounding boxes around objects. These bounding boxes can be stored easier since they need only the six bounding planes (in the simplest case). Due to the simplification of the object shapes, ray tracing can be performed faster since rays only have to be tested for whether they intersect with the object's bounding box. Only if a ray does intersect with a bounding box, the more geometrically complex objects within the bounding box are tested for intersection with the ray. To further increase performance, bounding boxes are stored in tree structures that allow more efficient ray traversal. For more on bounding volume hierarchies see [6].

2.3 Bounding Interval Hierarchy

The Bounding Interval Hierarchy data structure combines concepts of the Bounding Volume Hierarchy for storing objects and concepts of k-d trees for traversing the data structure. A big step in utilizing the Bounding Interval Hierarchy for fast ray tracing was achieved in 2006 by Carsten Wächter and Alexander Keller at the University of Ulm (see [2]). The Bounding Interval Hierarchy itself was first introduced by Beng Chin Ooi, Ron Sacks-Davis and K.

J. McDonell in 1987 (see [7]).

The Bounding Interval Hierarchy enhances the capacities of the Bounding Volume Hierarchy by reducing the number of stored splitting planes from six to two and at the same time increases flexibility compared to k-d trees as well as the Bounding Volume Hierarchy by allowing overlapping children and splitting planes that do not form an exact split of the parent box. This is accomplished by subdividing each stored bounding box along a split plane that is perpendicular to a chosen axis. Afterwards all objects are either sorted to the left or to the right side of that splitting plane. Additional to that, since there are two splitting planes for every subdivision there can be empty spaces between the resulting bounding boxes. This results in a further gain in performance when it comes to ray tracing because rays can now move through empty space where they are guaranteed not to intersect with any bounding planes (see figure 2). In order to choose the best splitting planes, the planes are chosen by a heuristic which results in more optimal trees. For more on the Bounding Interval Hierarchy see the work of Wächter and Keller [2].

2.4 Annotations

Notice that we assumed for all subdivision planes that they are axis-aligned. While it is generally possible to use non-axis aligned splitting planes in spatial subdivision data structures, throughout this paper when k-d trees or other similar data structures are used, the nodes stored will represent axis-aligned splitting planes.

3

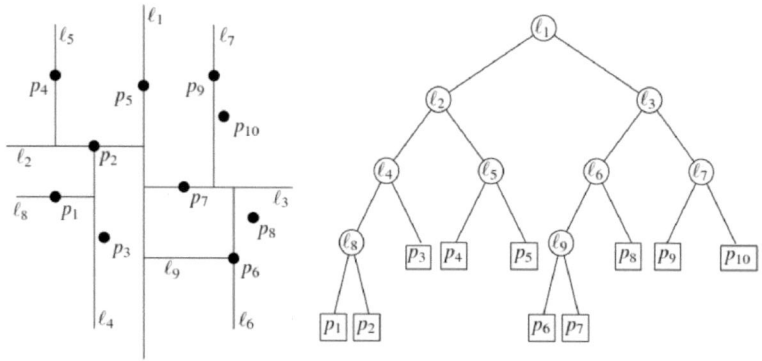

Figure 1: Partitioned Space and the resulting 1-dimensional k-d tree. The first splitting plane l_1 is put through p_5. The left and right children are the splitting planes through p_2 and p_7, l_2 and l_3. In this k-d tree not all points are intersecting a splitting plane, however they are still uniquely assigned to a leaf of the k-d tree. For the purpose of grid visualization, usually all points are stored as splitting planes. *Source:* [1]

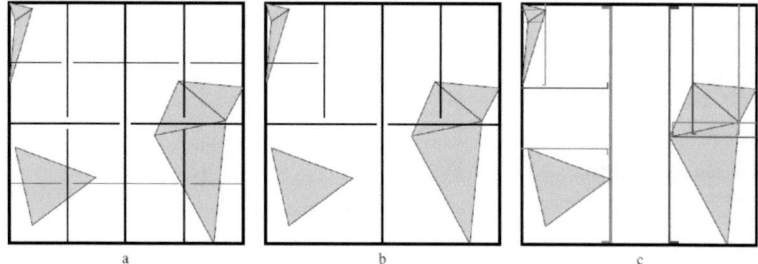

Figure 2: Three stages of constructing a Bounding Interval Hierarchy. In a) all possible split plane candidates are shown. In this case the candidates are determined by subdividing along the longest side of the axis-aligned bounding box. b) shows the split plane candidates that will actually be used. c) shows the resulting bounding interval hierarchy in which a heuristic chose the final splitting planes. In the image a red splitting plane implies a left child, a blue splitting plane implies a right child. *Source:* [2])

4

3 Point and Cell Location Methods

3.1 Efficient Point Location

At the time when Langbein et al. presented their paper, solutions to point location problems in unstructured grids were basically non-existent. The existing solutions for visualizing grids demanded for properties of the grids that were simply not matched in practice. Examples are preconditions like a uniform distribution of points, cell convexity or particular edge ratios. Their approach was to create an adaptive point-based k-d tree to divide space. Every leaf of the tree has an associated cell and contains only one vertex. In order to find a point this tree structure can be traversed to retrieve the index of the cell associated to the leaf that includes the point. The sought-for point is found by shooting a ray from the stored vertex to the point and tracing it by using cell adjacency information. If this takes place close to the boundary, problems can occur when the search ray intersects with the boundary. This case will be discussed in more detail in chapter 5.1.

3.2 Efficient Cell Location

With their work Garth and Joy overcame several shortcomings of Langbein et al.'s approach. For instance the data structure of Langbein et al.'s work relies heavily on cell adjacency information. When dealing with decomposed grids, this information can oftentimes not be stored and used properly and therefore Langbein et al.'s approach cannot be applied to such grids. The main difference of the work of

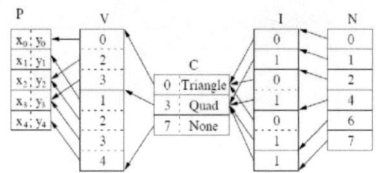

Figure 3: Data structure 1 and 2: Cell vertex information is stored in array P, V and C, adjacency information is stored in array I and N. *Source:* [4]

Garth and Joy is the different underlying data structure. They utilize the structure of a bounding interval hierarchy to create the *celltree* data structure. In order to locate a cell within the celltree, it is traversed to find a sufficiently small set of cell candidates that can then be tested for inclusion of the interpolation point in reasonable computing time.

4 Construction of Data Structures

4.1 k-d Tree and Adjacency Information

The data structure of Langbein et al. consists of a data structure containing a representation of the data set itself, a second data structure containing incidence information of cells and points and a third data structure representing the k-d tree.

As shown in figure 3, this data structure is represented by three arrays P, V and C. V and C are integer arrays. V stores all point indices incident to a cell, C stores cell type

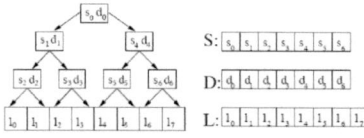

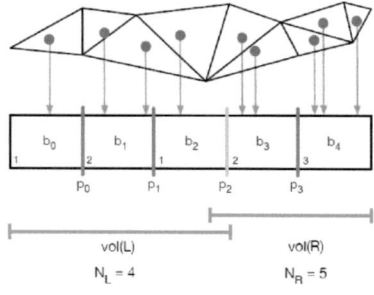

Figure 4: Data structure 3: k-d tree data structure. S stores split value, D stored the splitting dimension, L represents the leafs in form of an array of point indices. *Source:* [4]

Figure 5: Visualization of the split finding algorithm with equidistant buckets for $n_b = 5$. The bounding box centers of each cell determine their respective bucket. Afterwards the optimal splitting plane p_2 is determined in order to minimize later point inclusion test costs. *Source:* [3]

and offset into V. P is a float array and contains the coordinates of all points. There is an additional floating point array T not shown in figure 3 that stores the values for visualization in the form of 1,2,3,4,6 or 9 numbers for every point or cell.

Also shown in figure 3 is the storage of point to cell incidence information. The information is stored in the arrays N and I. In N the offset of the cell list stored in array I is stored for each point. I on the other hand is starting at the offset and stores all incident points for every cell.

Finally, the k-d tree is stored in three arrays S, D and L as shown in figure 4. We can distinguish between storage of inner nodes and storage of leafs. The inner nodes are stored in the arrays S and D. S is a floating point array that stores the split values, D is a character array that stores the split dimension. The leafs are stored in array L. L contains the point indices.

When constructing the k-d tree it is important to make good choices for the split dimension because boxes in the tree get thinner and thinner. In the work of Langbein et al. a mixture is used consisting of the natural choice of splitting along the largest

axis of each respective box and using the largest axis of the bounding box contained in the node.

The cell adjacency information is constructed by first counting the number of incident cells for each point and then subsequently traversing the resulting array in order to calculate the offsets for N. In the final two steps first all cells are traversed again and the number of incident cells is counted again in order to store the cell indices using the calculated offset, then the indices are sorted for every point.

4.2 The Celltree

Just as with the construction of the k-d tree, the choice of proper splitting planes is of fundamental importance for performance. In case of the celltree, the splitting planes are determined in order to create a tree that is as balanced as possible. The main reason for this is to keep the costs for cell inclusion tests low. There are different approaches to this such as splitting in the middle or splitting at the median. However Garth and Joy chose a different approach due to problems like strong imbalances or high traversal time with the split-middle and split-median approach. Instead, the idea is to span a set of n_b equidistant buckets over the entire bounding box of a list I of cell indices and sort every every index of I into one bucket by using the center of its bounding box as classification feature. After the cell indices have been sorted in buckets, a splitting plane is chosen to minimize the costs for later point inclusion tests. For more details on determining optimal splitting plane see [3].

5 Point and Cell Location

5.1 Cell Guessing and Ray Shooting in k-d trees

The idea for finding points within the k-d tree is to make an educated guess for a cell C_{old} close to the sought-for point and then shooting a ray from C_{old} towards the point. The reason for guessing a cell C_{old} first is that most of the time sought-for points are close to recently used cells, which is why the last used cell is taken as an initial guess for C_{old}. If there is no cell C_{old} or ray shooting fails because a boundary is hit or the distance of C_{old}'s center to the sought-for point is too large, the k-d tree is used. In this case the tree is traversed to find the leaf that contains the sought-for point. Rays are then cast from incident cells towards

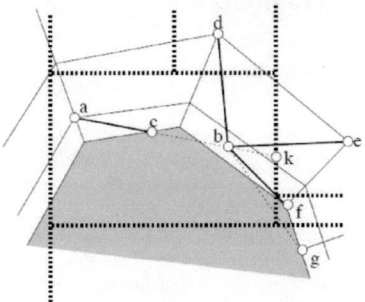

Figure 6: A leaf containing the sought-for point b also contains vertex a. A ray is shot from a towards b but hits the boundary in c. The elongation of the ray intersects with a leaf face in k. Using incidence information, d, e, f and g are identified as neighboring vertices and therefore suitable candidates for shooting further rays towards b. Here a ray shot from d finds the sought-for point b. *Source:* [4]

the vertex stored in the leaf. The point is found by tracing the rays from face to face (see figure 6).

5.2 Traversal of the Celltree and Point-in-Cell Test

The traversal and point-in-cell test of the celltree are implemented on a stack base starting at the root node. Once a leaf is reached it is checked for inclusion of the point. Garth and Joy used some minor tweaks for optimizing the traversal and point-in-cell test. For more on this see their work [3].

7

6 Performance

6.1 Methodology

Garth and Joy integrated their celltree data structure in two existing visualization frameworks. The first comparison is drawn with the Visualization Toolkit, or VTK. VTK is a commonly used framework for developing visualization applications. It has a class called *vtkCellLocator* which is built for cell location purposes. For testing purposes, Garth and Joy implemented a class *vtkCellTree* which was used as a substitute for the *vtkCellLocator* class. In order to obtain convincing results for the benchmarks, Garth and Joy did not replace VTK's inherent methods for cell interpolation and inclusion testing. Besides the *vtkCellLocator* class, VTK also contains a more modern cell locator called *vtkModifiedBSPTree*. It is a k-d tree based data structure that has an extra leaf for storing cells that overlap the splitting plane. Even though the *vtkModifiedBSPTree* class is designed for ray-casting applications it was included in the benchmark in order to obtain more representative results.

Further Garth and Joy compared their celltree with the adaptive k-d tree developed by Langbein et al. The k-d tree is implemented in the FAnToM visualization tool. FAnToM is not as widely used as VTK but is considered to be state-of-the-art when it comes to handling large unstructured grids. By comparing the celltree to a common solution like the VTK first, and then to an already strongly improved solution like FAnToM the remarkable advantage in performance and memory overhead will be more apparent. In order to benchmark the results, four visualization scenarios are run. First, a set of one million randomly distributed points are interpolated. Secondly a two-dimensional plane is chosen from the datasets challenging regions and points are interpolated on that grid. Thirdly, the same is done for a three-dimensional grid and lastly a set of streamlines is integrated into challenging regions of the grid. All benchmarks were run on a workstation with an Intel Core i7 2.66 GHz quad-core processor with 12GB of RAM. In section 4.2 it was shown that the points are sorted into buckets. The number of buckets does have an effect on performance. By empirical testing, Garth and Joy determined $n_b = 5$ results in good performance with acceptable build times. $n_b = 5$ was used for all benchmarks.

6.2 Datasets

As shown in figure 7 six datasets were chosen for benchmarking the celltree and the adaptive k-d tree:

- **Ellipsoid**
  This is an interesting dataset because on the one hand the distribution of cells is very uniform, on the other hand cell sizes vary greatly. The dataset is also reasonably small.

- **ICE and BMW**
  Cell location schemes generally have trouble dealing with very thin prism cells. These two datasets contain a large number of those thin prism cells so their benchmarking will presumably provide interesting results.

- **TDELTA**
  Around the edges of the sharp wing a large number of cells in the form of prisms is prevalent while at the same time the edges themselves present a

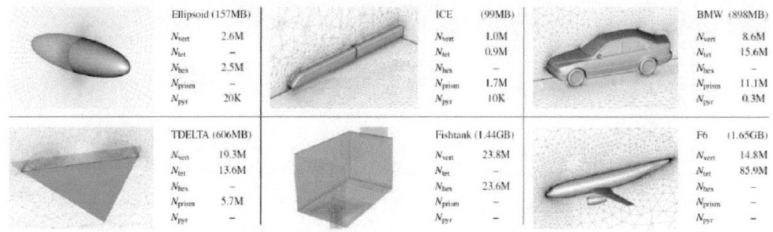

Figure 7: Overview over the six benchmarked datasets. *Source:* [3]

concave hole in the grid. These are two challenging factors that will hopefully result in interesting results.

- **Fishtank**
  The fishtank is the most interesting dataset. It is subdivided into 67,000 hexahedral subdomains, each of which is further subdivided in 9^3 rectangular structured grids. Due to local enumeration of the subdomains, the cells within one subdomain appear not to have any connectivity to cells of other subdomains.

- **F6**
  The F6 represents a very large grid consisting of almost 86 million tetrahedrons. Cell sizes vary greatly in size as the cells are very small close to the plane and grow larger quickly as we move further away from the plane.

6.3 Comparative Evaluation

The results of the benchmarks are shown in figure 8 and 9. The tables only include benchmarks that completed within ten minutes. The fishtank could not be benchmarked with FAnToM's built-in locator since, as described above, it lacks consistency in cell adjacency.

6.4 Celltree versus VTK

While build times of the *vtkCellTree* are higher by a factor of 5-8 compared to the *vtkCellLocator*, the celltree data structure is much more memory efficient by a factor of 4 to 5 when it comes to the smaller datasets and a factor of over 7 for the F6 with its many tetrahedrons. In terms of performance the results are fairly similar for the interpolation of random points, however for the plane, volume and streamline interpolation the celltree outperformed VTK's cell locator by a factor of 40-100, or, in the cases of the BMW, TDELTA and F6, VTK's built-in class did not even complete most of the benchmarks within ten minutes. The results of the *vtkModifiedBSPTree* class are similar, but for some benchmarks the performance of the modified k-d tree is significantly better than the performance of the *vtkCellLocator* on small datasets like the ellipsoid and the ICE. Yet, it is still outperformed by the celltree by a factor of 5-10 in most cases. In terms of memory overhead the celltree is more efficient than the *vtkModifiedBSPTree* by a

Dataset	Locator	Build time	Memory Overhead	Random	Plane	Volume	Streamlines
Ellipsoid	vtkCellTree	3.34s	22MB (8%)	4.32s	1.91s	11.58s	1.70s
	vtkCellLocator	0.61s	90MB (34%)	7.71s	75.68s	559.98	–
	vtkModifiedBSPTree	6.72s	236MB (91%)	30.37s	1.90s	116.33s	46.39s
ICE	vtkCellTree	3.27s	23MB (13%)	2.51s	1.68s	6.42s	2.66s
	vtkCellLocator	0.53s	115MB (65%)	5.95s	89.93s	57.65s	211.45s
	vtkModifiedBSPTree	7.65s	246MB (140%)	16.90s	13.43s	69.67s	43.57s
BMW	vtkCellTree	40.25s	229MB (14%)	2.34s	3.57s	8.94s	1.97s
	vtkCellLocator	5.01s	921MB (56%)	5.15s	–	–	–
	vtkModifiedBSPTree	303.23s	2476MB (151%)	–	–	–	–
TDELTA	vtkCellTree	28.08s	165MB (14%)	1.66s	4.65s	9.48s	25.33s
	vtkCellLocator	4.2s	880MB (79%)	3.99s	–	–	–
	vtkModifiedBSPTree	77.57s	1770MB (159%)	61.86s	–	–	–
Fishtank	vtkCellTree	27.59s	196MB (8%)	3.52s	3.79s	6.85s	27.11s
	vtkCellLocator	6.16s	851MB (35%)	7.74s	12.04s	40.04s	28.75s
	vtkModifiedBSPTree	199.41s	2162MB (91%)	–	–	–	–
F6	vtkCellTree	130.19s	743MB (16%)	1.59s	8.96s	17.63s	9.60s
	vtkCellLocator	22.40s	5426MB (124.54%)	5.80s	–	–	–
	vtkModifiedBSPTree	–	–	–	–	–	–

Figure 8: Benchmark results of the celltree embedded in VTK. *vtkCellLocator* is VTK's built-in cell location algorithm, the vtkCellTree is the drop-in replacement for *vtkCellLocator*. *Source:* [3]

Dataset	Locator	Build time	Memory Overhead	Random	Plane	Volume	Streamlines
Ellipsoid	FCellTree	5.19s	22MB (18%)	24.08s	2.28s	29.88s	0.83s
	FCellLocator	6.41s	150MB (76%)	21.17s	3.25s	30.14s	0.88s
ICE	FCellTree	4.93s	23MB (6%)	4.01s	1.93s	9.10s	3.22s
	FCellLocator	3.06s	88MB (85%)	11.87	37.81s	26.62s	4.32s
BMW	FCellTree	57.84s	229MB (10%)	7.91s	4.91s	21.36s	4.06s
	FCellLocator	24.56s	770MB (122%)	12.48s	51.88s	91.08s	4.01s
TDELTA	FCellTree	40.03	165MB (11%)	5.25s	7.33s	24.50s	5.11s
	FCellLocator	15.2s	483MB (82%)	11.84s	290.65s	–	6.33s
Fishtank	FCellTree	43.28s	196MB (5%)	14.37s	5.19s	24.89s	6.93s
	FCellLocator	–	–	–	–	–	–
F6	FCellTree	144.92s	734MB (56%)	2.55s	7.59s	25.72s	3.93s
	FCellLocator	40.66s	1633MB (81%)	12.04s	119.99s	132.23s	4.36s

Figure 9: Benchmark results of the celltree embedded in FAnToM. *FCellTree* is the drop-in replacement for FAnToM's built-in cell locator, *FCellLocator*. *FCellLocator* is an implementation of Langbein et al.'s approach for point visualization. *Source:* [3]

factor of 10 for all datasets.

6.5 Celltree versus FAnToM

I was especially interested in the results of this benchmark as it is a direct comparison of performance of the two approaches to cell location presented in this paper. Just as in the comparison with VTK, FAnToM's built-in class outperforms the celltree in build time. However again the celltree offers a significant advantage in memory overhead by a factor of 3-7. In terms of performance the *FCellLocator* class is very similar to the *FCellTree* when it comes to interpolating random points and especially streamlines, but in the plane and volume interpolations the celltree outmatches the k-d tree by factors of 2-40 and 2-5 respectively.

7 Further Research and Optimization

Langbein et al. mention possible shortcomings of their k-d tree based approach when the grid exhibits ragged borders with low cell resolution. The preprocessing for such grids is very time consuming which is why they propose to further research this scenario. However, with the work of Garth and Joy this problem can be considered overcome. Nevertheless, the celltree structure as presented in this paper does leave plenty of room for further research. The most apparent starting point is to utilize the power of GPUs to further speed up interpolation of points. Garth and Joy took first steps towards this already as shown in section 7.1.

Besides the GPU interpolation, implement-ing the celltree in the OpenGL shading language constitutes a promising starting point for further research in order to enable unstructured interpolation directly from the rendering phase of visualization applications. Further, the celltree may be employed in the OpenCL framework. Finally, utilizing parallel computing in order to enable distributed interpolation over decomposed unstructured grids on clusters could be a very promising starting point. However, I could not find any publications of neither the Computational Topology Group of the University of Kaiserslautern nor of the IDAV Institute for Data Analysis and Visualization of the University of Davis, California. IDAV did present results for parallelization algorithms (for example [8]), however they targeted structured grids. Nevertheless, the work presented in [8] was implemented in CUDA just like the attempts of Garth and Joy to use GPUs, therefore it might be possible to draw considerable conclusions from [8].

7.1 GPU Interpolation

Given its inherent data structure, generally the celltree allows for utilization of GPUs. To determine the potential of speed gain through GPU interpolation for the celltree Garth and Joy implemented it in the CUDA programming language. Generally unstructured grids present a real challenge for GPUs since they contain a number of indirections that does not work well with GPU architectures. To overcome this problem, Garth and Joy came up with some tweaks for the internal representation of the celltree to be better-suited to GPU architectures. In order to benchmark the GPU-approach they implemented a particle advection scheme as shown in figure 11. The

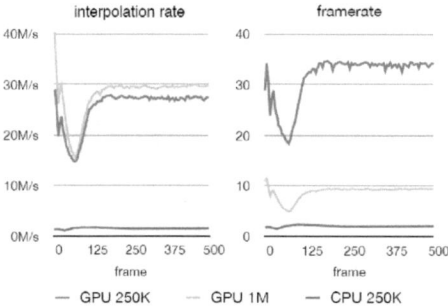

Figure 10: Interpolation and frame rates for the GPU benchmark. *Source:* [3]

Figure 11: Visualization of one million particles advecting into a fishtank. *Source:* [3]

test was run on a NVidia GeForce 285GTX with 2GB of VRAM. To sum up the results of figure 10, the benchmark showed an average speedup by the factor of 16.5 for the test data. At the same time the memory overhead caused by the less efficient GPU-compatible grid storage was about 8% and therefore almost neglectable.

The GPU implementation presented here was just a first step in utilizing the processing power of GPUs. This is due to the fact that while Garth and Joy used some tweaks to make the data structure representation suitable for GPU architectures, they did not optimize any build parameters yet. Further the representation can be optimized in regards to the respective GPU architecture and there is great optimization potential for the routines for point-in-cell testing as they account for 90% of the register and instruction count.

8 Summary

In this paper I presented two approaches for point visualization in large, unstructured grids. First the k-d tree based approach of Langbein et al. was introduced, second the celltree based on a bounding volume hierarchy by Garth and Joy was presented. Langbein et al. use an adaptive k-d tree to divide space, associate a cell with every leaf of the tree and use ray shooting to locate points in the grid using adjacency information. Garth and Joy use a bounding volume hierarchy to store the grid and use it to narrow down cell candidates which are afterwards tested for point inclusion to determine a points location in the grid. The work of Garth and Joy was presented more than seven years after the approach of Langbein et al. and can be considered an enhancement of their work. In a direct comparison, the celltree outperformed the k-d tree by a factor of 3-7 in memory overhead and up to a factor of 40 in speed with the average factor being between 2 and 5. Further research could be conducted on the possibilities of maximizing the potential of GPU processing power and parallel distributed interpolation on clusters.

12

References

[1] M. de Berg, O. Cheong, M. van Kreveld, M. Overmars: *Computational Geometry: Algorithms and Applications.* 3rd edition, Springer, 2008.

[2] C. Wächter, A. Keller: *Instant Ray Tracing: The Bounding Interval Hierarchy.* In Eurographics Symposium on Rendering, 2006.

[3] C. Garth, K.I. Joy: *Fast, Memory-Efficient Cell Location in Unstructured Grids for Visualization.* In Visualization and Computer Graphics, IEEE Transactions on, vol.16, no.6, pp.1541-1550, Nov.-Dec. 2010.

[4] M. Langbein, G. Scheuermann, X. Tricoche: *An efficient point location method for visualization in large unstructured grids.* In Proceedings of Vision, Modeling, Visualization, 2003.

[5] J. L. Bentley: *Multidimensional binary search trees used for associative searching.* Communications of the ACM, 18(9):509-517, 1975.

[6] H. J. Haverkort: *Results on Geometric Networks and Data Structures.* Ph.D. thesis, Utrecht University, 2004, pp.5-21.

[7] B. C. Ooi, K. J. McDonell, R. Sacks-Davis: *Spatial kd-tree: An indexing mechanism for spatial databases.* In Proceedings of IEEE Int. Comp. Software & Applications Conf., Japan, 1987.

[8] Christopher P. Stone, Earl P. N. Duque, Yao Zhang, David Car, John D. Owens, Roger L. Davis: *GPGPU parallel algorithms for structured-grid CFD codes.* In Proceedings of the 20th AIAA Computational Fluid Dynamics Conference, no. 2011-3221, 2011.

13

YOUR KNOWLEDGE HAS VALUE